REGIONAL TECHNICAL COLLEGE
LETTERKENNY

This book is due for return on or before the last date shown below.

Don Gresswell Ltd., London, N.21 Cat. No. 1207

Optical
Distance
Measurement

Optical Distance Measurement

D. J. HODGES, B.Sc., Ph.D., C.Eng.,
F.I.Min.E., M.Inst.F., F.G.S.
(*Reader in Surveying and
Mining Engineering, University
of Nottingham*)

and

J. B. GREENWOOD, B.Sc., M.Sc.
(*Mining Engineer, Anglo-American
Mining Corporation of South Africa*)

LONDON

BUTTERWORTHS

THE BUTTERWORTH GROUP

ENGLAND

Butterworth & Co. (Publishers) Ltd.
London: 88 Kingsway WC2B 6AB

AUSTRALIA

Butterworth & Co. (Australia) Ltd.
Sydney: 20 Loftus Street
Melbourne: 343 Little Collins Street
Brisbane: 240 Queen Street

CANADA

Butterworth & Co. (Canada) Ltd.
Toronto: 14 Curity Avenue, 374

NEW ZEALAND

Butterworth & Co. (New Zealand) Ltd.
Wellington: 49/51 Ballance Street
Auckland: 35 High Street

SOUTH AFRICA

Butterworth & Co. (South Africa) Ltd.
Durban: 33/35 Beach Grove

First published 1971

ISBN 0 408 70092 0

Filmset and printed in England by
Cox & Wyman Ltd., London, Fakenham and Reading

Contents

Preface

In a decade which has seen the mapping of the surface of the moon by lunar orbiting satellites, perhaps man's most spectacular surveying programme so far, it is not surprising that many important developments in surveying instruments and techniques have taken place. The increasing use and variety of surveying instruments available, particularly for distance measuring, has resulted in a need for textbooks dealing specifically with these developments.

This book deals with the instruments and techniques used for measuring distances by optical means. It is based on a series of articles and lecture notes prepared by the authors over the past few years, and embodies the authors' experience in teaching the subject. The theory and principles of each system are explained in detail and at least one typical instrument from each system is fully described. The accuracies obtainable and the errors influencing the various methods are discussed, and practical hints on their use in the field given.

Modern optical distance measuring instruments enable true horizontal and vertical distances to be measured directly, without the necessity of computations or mathematical tables. Horizontal and vertical angles can also be measured with these instruments, and as their overall size and weight is similar to conventional surveying instruments, they fulfil the functions of a good theodolite.

It is hoped that the book will not only be of use to practising surveyors and engineers but also to teachers and students, there being no equivalent text in the English language which covers the subject in such detail.

The authors would also like to express their sincere gratitude to all the firms, organisations, and other persons who have assisted them in the preparation of this book.

D. J. Hodges
J. B. Greenwood

CHAPTER 1

Introduction and Brief Historical Review

During the past decade there has been an increasing demand for survey instruments which enable distances to be measured quickly and accurately by optical methods, particularly instruments which will determine horizontal and vertical distances directly without the necessity of reduction tables or lengthy calculations.

The main advantage of all optical methods is that poor surface conditions do not have any adverse effects on accuracy and, in certain instances, measurements can be obtained which would otherwise have been impossible with normal taping methods. Another advantage is the ease and convenience of controlling the accuracy to suit the requirements for any particular survey task. Since the error is usually directly proportional to the distance measured, a line may be divided into small sections when great accuracy is required and into longer sections when accuracy is less important and greater speed required.

The surveyor now has the choice of many methods and instruments for optical distance measuring and, by the selection of the appropriate instrument and method, accuracies to satisfy a wide range of measurements can be attained.

High efficiency and economy may be obtained with vertical staff methods, particularly in areas in which many points need to be surveyed and a high accuracy is not of paramount importance (up to about $\frac{1}{2000}$). Such methods are used in the preparation of plans for the construction of roads and railways, and in topographical surveys which are intended for making maps for educational uses and government services. They can be used to great advantage to supplement detail on maps produced by photogrammetric techniques, particularly to provide information over small areas which may be obscure on photographs.

1

Due mainly to differential refraction, vertical staff methods do not permit the accuracy attainable with subtense bar and horizontal staff methods, which in many instances can be superior to good tape measurements (up to about $\frac{1}{10000}$). On the other hand, the time required to set up a vertical staff will generally be much less than that required for a horizontal staff or subtense bar. However, a combination of both systems can be included in the same survey, for example, in a traverse the polygon sides may be measured with horizontal staves and the detail fixed by vertical staff observations. In this way the surveyor may utilise the advantages of both systems.

This is particularly advantageous when instruments such as the Fennel FLT Code-theodolite and the Kern Code-Tacheometer are available and used in conjunction with electronic computers and automatic plotting machines. Methods can be applied which otherwise would have been considered too complex for conventional means of computation although requiring comparatively simple field work. These instruments and methods represent a major advance in the development of fully automated survey methods.

During the past century numerous instruments have been developed for optical distance measuring, many of which have only an historical value. Nevertheless, a brief review of the development of the more important types is given below.

Although the main principles of certain instruments, familiar to the authors, are described, this is not intended to decry the merits of other makes. It should also be noted that although this textbook deals primarily with optical distance measurement, many of the instruments described will also measure angles to a few seconds of arc, and therefore provide a means of observing both angular and linear measurements from the same instrument station. Thus the surveyor can have complete and constant control of all facets of the work and, in addition to greater convenience and economy, this should reduce the number of gross errors.

1.1   THE DEVELOPMENT OF OPTICAL DISTANCE MEASUREMENT

The development of instruments and techniques for indirect distance measurement has been taking place for many centuries. It is difficult to pinpoint the exact date and place of the earliest attempts at optical distance measurement, since records show that similar developments were being carried out in several countries at approximately the same time.

The English astronomer, Gascoigne, was probably the first to use the subtense bar method of measurement, whilst an instrument employing a stadia micrometer was designed by the Italian, Montanari, a Venetian doctor, about 1674. James Watt is thought to have used a similar type of instrument to Montanari in his survey of the West of Scotland in 1774, and further improvements were made in England by Green about 1778. A stadia micrometer tacheometer was also being manufactured about this time by M. de la Hire in France.

In Germany, the fixed theodolite stadia wires were introduced and used in Augsburg by Brander in 1764, and, during the first Bavarian land survey, theodolite tacheometry was introduced for the detail measurement. As early as 1830 the Italian, Ignazio Porro, invented the anallactic telescope by the insertion of a supplementary lens between the objective and the diaphragm cell, thereby simplifying the reduction equations. The gradual improvement of the theodolite telescope for tacheometric measurements has continued up to the present time.[1, 2]

The oblique distances obtained with fixed stadia theodolite tacheometry must be mathematically reduced to the horizontal and, for many tasks, such measurements are not sufficiently accurate. These disadvantages led to the development of various designs of self-reduction tacheometers, such as projection tacheometers, contact tacheometers, diagram circle tacheometers, double-image tacheometers, and movable diaphragm wire tacheometers. The development of the handier diagram circle instruments has practically eliminated the use of the projection and contact types of instrument.

Tacheometers with movable diaphragm lines such as the Kern DK–RV and K1–RA models are constructively specially attractive, and are becoming increasingly acceptable in general practice. The Jeffcott tacheometer,[3] patented in England in 1912 and formerly manufactured by Cooke, Troughton and Simms, was of the movable diaphragm line type and probably one of the first truly self-reducing instruments made. At the present time, none of the British survey instrument firms manufacture a self-reduction tacheometer, although Hilger and Watts have a tacheometric prism for mounting onto their microptic theodolites.

The construction of the diagram circle tacheometer is associated with Ernst von Hammer and Adolf Fennel.[4] The first of these instruments built by Fennel, using the ideas of Hammer, appeared at the turn of the century. Further developments were made by Leeman[5] in Switzerland, and Dahl[6] in Norway. Examples of this design are manufactured by several firms, notably Zeiss (the 'Dahlta' 010), Wild (the RDS), Fennel (the F.T.R.A.), Metrimpex (the TA–DI) and Salmoiraghi (the Model 4180). Kern no longer manufacture this type of instrument, their type DKR diagram circle tacheometer being replaced by the type DK–RV, but diagram circles are still used in their self-reducing plane table alidades.

Hammer received his first inspiration for the construction of a diagram circle tacheometer from a thesis by the Italian engineers, G. Roncagli and E. Urbani ('Theory and Description of the Reduction Tacheometer'). The instrument described in this thesis possessed a diagram with distance measurement bands whose manual positioning followed the earlier reading of the vertical angle. Hammer extended the diagram by the addition of a double height curve and he developed a method of automatically positioning the diagram with the tilting of the telescope. Hammer published his ideas in 1898 and thereby found Fennel, who promoted and worked on the detailed construction of the instrument. The results of this Hammer–Fennel partnership were so successful that by 1900 the first trial measurements could be undertaken. Since then, the diagram circle tacheometer has improved continuously and substantially along with the improvements in the design of theodolite optical systems generally, notably the shortened telescope and internal focussing.

In the original Hammer–Fennel tacheometer the image of the diagram covers up one-half of the field of view of the telescope. The staff could not, therefore, be lined up centrally, and the small size of the diagrams, with steep altitude curves, resulted in difficulty in estimating the intersection point between the curve and the staff. To overcome these difficulties, the Swiss, Leeman, developed, in 1933, an instrument in which the Hammer diagram was marked onto a glass circle situated immediately in front of the eyepiece, and which rotated with the tilting of the telescope. The telescope field of view was wholly usable and unrestricted.

Following the suggestions of the Norwegian engineer Dahl, Zeiss built a new tacheometer in 1942 in which the diagram curves were marked directly onto the vertical circle. Again the whole of the telescope field of view was available and the vertical circle was also read in the telescope eyepiece. This instrument was further improved in 1952 and is now manufactured as the Zeiss 'Dahlta' 010. The tacheometer built by the firm Wild under the name RDS appeared in 1950. This differs from the Zeiss instrument in that the diagram circle is not fixed but is connected to the telescope so that it turns with the tilting of the telescope but in the opposite direction. This results in gently sloping height curves which increase the ease and accuracy of staff reading.

A completely different and ingenious approach to the development of a surveying instrument, which would enable distance to be measured directly, was proposed by R. Bosshardt[7] about 1920. Again it was the Carl Zeiss works who constructively realised the ideas of Bosshardt with the 'Redta' 002 self-reducing double-image tacheometer. The latest instruments of this type are the Kern DK–RT, Zeiss 'Redta' 002 and the Wild RDH all based on the Bosshardt system. Sometimes called 'precision' tacheometers, these instruments are as accurate as steel-band measurements over horizontal ground and have a superior accuracy over undulating terrain. They will also measure angles to a few seconds of arc.

A number of survey instruments are used by military authorities, a well-known example being the rangefinder. The modern single-observer rangefinder was invented by Barr and Stroud in the year 1888. Various modifications have been introduced over the years and present-day models are considerably more refined in every respect. The accuracies now obtainable in both direction and distance measurements permit the application of the rangefinder to general traverse work.

1.2 RECENT DEVELOPMENTS

Soon after the introduction of electronic data processing techniques, the digital computer was employed for the solution and plotting of field readings. Further advances have since been made and the entire surveying procedure has been completely automated. Code theodolites and tacheometers can be used in the field and the results from such instruments are translated from their coded form, which is recorded on 35 mm film, into punched tape which can be translated by electronic equipment. Hence, human errors both in the field and in the office have been eliminated from the traditional surveying practice. Also the amount of labour and time involved has been drastically reduced.

The present-day techniques for optical distance measurement can be seen to be fundamentally the same as those used by a number of undoubtedly brilliant pioneers. The instruments, however, have changed considerably, now being generally lighter, more accurate and mechanically and optically more refined in every respect.

REFERENCES

1. TAYLOR. 'A new perfectly anallactic internally focussing telescope.' *Trans. Opt. Soc.* **25**, 4 (1923–24).
2. TAYLOR and HOGG. 'Applications of the Surveyor's telescope to precise tacheometry.' *Paper given at Commonwealth Survey Conference* (1947).
3. JEFFCOTT. 'A direct-reading tacheometer.' *Trans. Inst. C.E.I.*, Vol. XLI.
4. HAMMER. 'Der Hammer-Fennelsche Tachymeter-Theodolit zur unmittelbarn Lattenablesung von Horizontaldistanz und Hohenuntershied.' *Zeitschr. f. Instrumentenkunde,* **22**, S.21 (1902).
5. LEEMAN. 'Uber eine neue selbstreduzierende Kippregel der Firma Kern.' *Schweizer Zeitschr. f. Vermussungs,* S.56, K (1936).
6. WEKMEISTER. 'Reduktionstachymeter, Dahlta.' *Zeitschr. f. Vermussungs,* S.160, (1942).
7. BOSSHARDT. *Optische Distanzmessung und Polar-koordinaten methode.* Witwer Stuttgart (1930).

CHAPTER 2

Theodolite Tacheometry

2.1 INTRODUCTION

All optical methods of determining distance are based on the formation of an acute-angled triangle which can be solved by various combinations of base and subtense angle measurements.[1]

In subtense bar measurements, the base length is constant, and the subtended angle is measured, whereas in rangefinding instruments the base is normally situated at the instrument station and can be either of constant or variable length. The base can be vertical or horizontal, and the instruments self-reducing giving true horizontal and vertical distances directly.

In theodolite tacheometers with a fixed parallactic angle, α, the staff intercept or base length is proportional to the distance; this is the stadia intercept method. The triangle may also be solved by tangential methods.

2.2 STADIA INTERCEPT METHODS

Most modern theodolites have diaphragm cells with two extra parallel stadia lines equally spaced on either side of the central horizontal line. If a staff is sighted through the telescope of such an instrument, and the readings of the outer lines noted, the difference in the readings, known as the staff intercept, will be directly proportional to the horizontal and vertical distances between the instrument and staff. With horizontal sights the horizontal distance will be obtained immediately and engineers' levels fitted with horizontal circles may be used. If the sights are inclined, slope distances are obtained and these must be reduced to the horizontal by computation, and the observations must be taken with a theodolite. The theodolite telescope may be specially designed for tacheometric

7

observations. It usually has a high magnification and contains lenses which ensure that distances are measured to the trunnion axis of the instrument.

Generally, the distance between the two extra stadia lines is designed in such a way that the distance between the instrument and staff is given by multiplying the staff intercept by a constant, usually 100. The distance between the stadia lines remains fixed for all values of inclination of the telescope and for all distances, and thus this method is normally referred to as the 'Fixed Stadia Method'.

2.2.1   FIXED STADIA METHOD

With this method there are three cases to be considered:

1. The special case of horizontal sights.
2. Inclined sights with staff held vertically.
3. Inclined sights with staff normal to the line of sight.

2.2.1.1   *Horizontal sights*

Fig. 2.1 shows the paths of rays of light travelling from the staff and through the objective lens of the theodolite telescope. With reference to this figure, we have the following

Triangles x*O*y and *XOY* are similar.

$$\therefore \qquad \frac{v}{u} = \frac{s}{i} \qquad (2.1)$$

From the elementary lens formula:

$$\frac{1}{f} = \frac{1}{u} + \frac{1}{v} \qquad (2.2)$$

where u and v are the conjugate focal distances and f is the focal length of the lens.

Multiplying through by fv

$$v = f\frac{v}{u} + f \qquad (2.3)$$

Substituting from equation (2.1)

$$v = f\frac{s}{i} + f \qquad (2.4)$$

Adding c to both sides

$$v + c = f\frac{s}{i} + f + c \qquad (2.5)$$

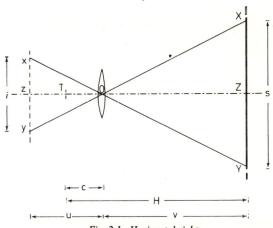

Fig. 2.1. Horizontal sights

u. *Distance from objective to image*
v. *Distance from objective to staff*
O. *Centre of objective lens*
T. *Trunnion axis of theodolite*
i. *Stadia interval (height of image)*
c. *Distance from trunnion axis to objective*
H. *Horizontal distance between instrument and staff stations*
s. *Staff intercept*

but from *Fig. 2.1*

$$v + c = H$$

∴

$$H = f\frac{s}{i} + (f + c) \qquad (2.6)$$

Replacing $\frac{f}{i}$ by K_m and $(f + c)$ by K_a we have

$$H = K_m s + K_a \qquad (2.7)$$

where

K_m is the multiplying constant
K_a is the additive constant

2.2.1.2   Field determination of the multiplying constant (K_m) and the additive constant (K_a) of a theodolite

The method for the determination of the multiplying and additive constants in the field is to set out a nearly level line of say 200 m with intermediate stations at 20 m intervals. The instrument is then set up at one end of the base and stadia readings are taken successively onto a staff held vertically on each of the stations in turn (*Fig. 2.2*).

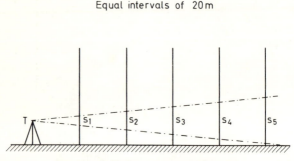

Equal intervals of 20 m

Fig. 2.2. *Field determination of the multiplying constant (K_m) and the additive constant (K_a) of a theodolite*

T. *Trunnion axis of theodolite*
$s_1, s_2, s_3 \ldots s_n$. *Staff intercepts at known distances*

The constants can then be determined by substituting any selected pair of observations into equation (2.7) and by solving the simultaneous equations so obtained.

The constants can also be determined by plotting the graph of staff intercepts ($s_1, s_2, s_3 \ldots\ldots s_n$) as abscissae, and the known values of the corresponding horizontal distances (H) as ordinates. The constants K_m and K_a can be determined from the slope and intercept respectively of the resulting straight line graph. If a precise determination is required the equation of the line can be obtained by the Method of Least Squares[2] so that $\sum(H - K_m s - K_a)^2$ is a minimum for the optimum values of the two constants. In most modern theodolites and levels the values of K_m and K_a are 100 and 0 respectively.

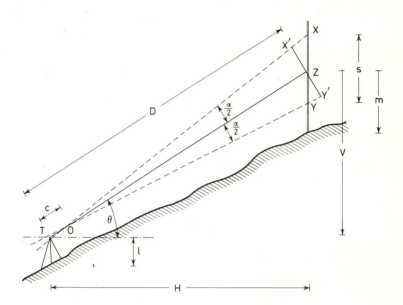

Fig. 2.3. Inclined sights with staff held vertically

D. *Slope distance from instrument trunnion axis to central stadia line intercept on staff*

H. *Horizontal distance from instrument trunnion axis to staff*

V. *Vertical distance from instrument trunnion axis to central stadia line intercept on staff*

c. *Distance from trunnion axis to objective*

θ. *Angle of elevation of telescope*

s. *Staff intercept*

m. *Height of central stadia line intercept above base of staff*

l. *Height of instrument trunnion axis above the ground*

$\frac{\alpha}{2}$. *Angle subtended at the instrument by one stadia line and line of sight*

O. *Objective lens of telescope*

T. *Trunnion axis of theodolite*

X, Z, Y.   *Points intercepted by stadia lines on staff*

X', Z, Y'. *Projection of X, Z, Y onto line normal to line of sight*

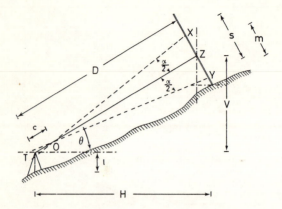

Fig. 2.4. *Inclined sights with staff normal to line of sight*

D. *Slope distance from instrument trunnion axis to central stadia line intercept on staff*

H. *Horizontal distance from instrument trunnion axis to staff*

V. *Vertical distance from instrument trunnion axis to central stadia line intercept on staff*

c. *Distance from trunnion axis to objective*

θ. *Angle of elevation of telescope*

s. *Staff intercept*

m. *Height of central stadia line intercept above base of staff measured along the staff*

l. *Height of instrument trunnion axis above the ground*

$\frac{\alpha}{2}$. *Angle subtended at the instrument by one stadia line and line of sight*

O. *Objective lens of telescope*

T. *Trunnion axis of theodolite*

X, Z, Y. *Points intercepted by stadia lines on staff*

2.2.1.3   *Inclined sights with staff held vertically*

The more general case of inclined sights with the staff held vertically
is shown in *Fig. 2.3*.

As the staff is not normal to the line of sight we cannot apply
equation (2.6) directly, but considering the line $X'ZY'$ perpendicular
to the line of sight we have

$$\text{angle } XZX' = \theta$$
and $$\text{angle } XX'Z \simeq 90°$$

and we may assume $X'Y' = s \cos \theta$ with negligible resulting error.
Hence, comparing with equation (2.7), we have

$$D = K_m s \cos \theta + K_a \qquad (2.8)$$

and $$H = D \cos \theta$$

$\therefore$ $$H = K_m s \cos^2 \theta + K_a \cos \theta \qquad (2.9)$$

similarly for V we have

$$V = K_m s \cos \theta \sin \theta + K_a \sin \theta \qquad (2.10)$$

It must, however, be noted that the true difference in ground level
between the two stations is

$$V - m + l$$

Hence, the true difference in level

$$= K_m s \cos \theta \sin \theta + K_a \sin \theta - m + l \qquad (2.11)$$

When the sights are depressed this becomes

$$K_m s \cos \theta \sin \theta + K_a \sin \theta + m - l \qquad (2.12)$$

Hence from equations (2.9), (2.11) and (2.12) the true horizontal
and vertical displacements between the two stations can be readily
calculated.

2.2.1.4   *Inclined sights with staff normal to line of sight*

With reference to *Fig. 2.4* since the staff is perpendicular to the line
of sight, we may write

$$D = K_m s + K_a$$

and since

$$H = D \cos \theta + m \sin \theta$$

we have

$$H = K_m s \cos \theta + K_a \cos \theta + m \sin \theta \qquad (2.13)$$

For depressed sights

$$H = K_m s \cos \theta + K_a \cos \theta - m \sin \theta \qquad (2.14)$$

For the vertical displacement, V, we have

$$V = D \sin \theta$$

∴ $$V = K_m s \sin \theta + K_a \sin \theta \qquad (2.15)$$

The true difference in level between the two stations is given by

$$K_m s \sin \theta + K_a \sin \theta - m \cos \theta + l \qquad (2.16)$$

for elevated sights and

$$K_m s \sin \theta + K_a \sin \theta + m \cos \theta - l \qquad (2.17)$$

for depressed sights.

2.2.2 MOVABLE STADIA METHOD

In this method the stadia lines are displaced by a micrometer screw over a fixed staff intercept. Equation (2.6) can be written in the form

$$H = r k + (f + c)$$

where k is a constant proportional to the product of the focal length of the lens system and the staff intercept ($f \times s$) and r a measured quantity, inversely proportional to a variable value of the stadia interval, read direct on the micrometer. The method possesses the advantage that the stadia line can be moved to coincide with a definite staff graduation (as in precise levelling when using a parallel plate micrometer) and hence interpolation errors are minimised. The method, which is not much used, is similar to the subtense method using the stadia principle.

2.2.3 THEORY OF THE ANALLACTIC LENS

The additive constant $(f+c)$ occurs repeatedly in the foregoing analysis and this factor causes considerable complications in the resulting computations, especially in older theodolites with external focussing in which the distance c varies with the distance of the staff.

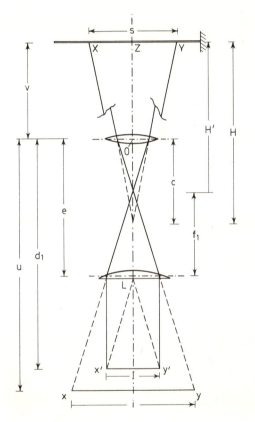

Fig. 2.5. Theory of the anallactic lens

The solution of this problem, by the elimination of the additive constant, is credited to an Italian geodesist, Ignazio Porro, sometime about 1830. He suggested a modified telescope system comprising an anallactic lens mounted behind the objective lens and at a constant distance from it.

From equation (2.6)

$$H = s\frac{f}{i} + (f+c)$$

it can be seen that the staff intercept (s) is directly proportional to $H-(f+c)$, that is, the distance between the staff station and the exterior principal focus of the objective lens. The principal focus (as seen in *Fig. 2.5*) therefore forms the apex of a constant visual angle, the extremities of which intercept the value of s on the staff. Thus if the apex of this angle was situated at the vertical axis of the instrument, the ($f+c$) term would vanish and the horizontal distance (H) would then be directly proportional to the staff intercept (s). It is possible to accomplish this by the introduction of an additional lens known as the anallactic or Porro lens into the telescope.[1, 3]

With reference to *Fig. 2.5*, O and L are the optical centres of the objective and anallactic lenses respectively, x and y represent the extremities of the image produced *before* the introduction of the anallactic lens, and x' and y' the extremities of the new image formed *after* the inclusion of the lens.

Let H = horizontal distance of the staff to the instrument trunnion axis

$\quad\quad v$ = horizontal distance of the objective lens to the staff

$\quad\quad f$ = focal length of the objective lens

$\quad\quad c$ = horizontal distance from trunnion axis to centre of objective lens

$\quad\quad u$ = distance from optical centre of objective lens to image xy

$\quad\quad d_1$ = distance from the optical centre of the objective lens to the actual image $x'y'$

$\quad\quad f_1$ = focal length of the anallactic lens

$\quad\quad e$ = distance between optical centres of objective and anallactic lenses

$\quad\quad i$ = length xy

$\quad\quad I$ = length $x'y'$, the actual stadia interval.

From the elementary lens formula for the image xy we have

$$\frac{1}{f} = \frac{1}{u} + \frac{1}{v} \tag{2.18}$$

and

$$\frac{v}{u} = \frac{s}{i} \tag{2.19}$$

Now the image xy acts as a virtual object for the newly introduced anallactic lens. Hence

$$\frac{1}{f_1} = \frac{1}{(d_1 - e)} - \frac{1}{(u - e)} \qquad (2.20)$$

(The negative sign is necessary to account for the virtual object in keeping with the 'real is positive, virtual is negative' sign convention.) Also

$$\frac{i}{I} = \frac{(u - e)}{(d_1 - e)} \qquad (2.21)$$

From these equations, we can arrive at an expression for H, bearing in mind the relationship

$$H = v + c \qquad (2.22)$$

From equation (2.20)

$$(d_1 - e) = \frac{f_1(u - e)}{f_1 + u - e} \qquad (2.23)$$

From equation (2.21)

$$i = \frac{I(u - e)}{(d_1 - e)} \qquad (2.24)$$

Combining equations (2.23) and (2.24) we get

$$i = \frac{I(f_1 + u - e)}{f_1} \qquad (2.25)$$

but from equation (2.18)

$$u = \frac{fv}{v - f} \qquad (2.26)$$

$\therefore$ substituting for u in equation (2.25)

$$i = \frac{I\left(f_1 + \dfrac{fv}{(v-f)} - e\right)}{f_1} \qquad (2.27)$$

From equation (2.19)

$$i = s\frac{u}{v} \qquad (2.28)$$

From equations (2.26) and (2.28)

$$i = \frac{sf}{(v-f)} \tag{2.29}$$

Thus equating (2.27) and (2.29)

$$i = \frac{I\left(f_1 + \dfrac{fv}{(v-f)} - e\right)}{f_1} = \frac{sf}{(v-f)} \tag{2.30}$$

From equation (2.30)

$$v = \frac{ff_1 s}{I(f_1 + f - e)} - \frac{f(e-f_1)}{(f+f_1-e)} \tag{2.31}$$

From equations (2.22) and (2.31)

$$H = \frac{ff_1 s}{I(f_1 + f - e)} - \frac{f(e-f_1)}{(f+f_1-e)} + c \tag{2.32}$$

Hence for H to be directly proportional to s we require the terms

$$-\frac{f(e-f_1)}{(f+f_1-e)} + c \text{ to disappear}$$

i.e.

$$c = \frac{f(e-f_1)}{(f+f_1-e)} \tag{2.33}$$

Here we have a relationship between c, the two focal lengths (f and f_1) and e, the distance between O and L which can be varied by design.

From equation (2.33)

$$e = f_1 + \frac{fc}{f+c} \tag{2.34}$$

With e chosen as above, the required effect is realised, namely that the apex of the visual angle is now situated at the vertical axis of the telescope.

Equation (2.32) can then be written

$$H = K_m s$$

Where

$$K_m = \frac{ff_1}{I(f_1 + f - e)} \tag{2.35}$$

The factor K_m is usually 100, so that $H = 100s$.

Theoretically, a truly anallactic telescope cannot be obtained by the addition of a single simple lens only, because the internal focussing lens moves with respect to the objective and therefore the effective focal length is thus variable, whereas the distance between the anallactic lens and objective is fixed.

Most modern anallactic telescopes have two or more supplementary lenses instead of one between the objective lens and the diaphragm cell.

The problem of the loss of light caused by the introduction of the extra lenses has been overcome by selective 'blooming' of the optical elements and by increasing the aperture of the objective.[4]

2.3 THE TANGENTIAL METHOD

In the foregoing analysis, the intercept subtended on the staff by two fixed or movable stadia lines has given a measure of the horizontal distance (H) between instrument and staff stations. With the tangential method no extra fittings are required on the theodolite, but two separate pointings of the telescope are necessary. As the stadia lines play no part in this method, the anallactic properties of the telescope do not enter into the computations.

2.3.1 HORIZONTAL SIGHTS

Assuming the telescope to be horizontal, the staff reading at Y is noted and the telescope is then raised through a measured vertical angle θ and the reading at X is recorded (see *Fig. 2.6*). Subtraction of the two readings will give the staff intercept (s) and knowing θ we have

$$\tan \theta = \frac{s}{H}$$

$\therefore \qquad\qquad H = s \tan \theta \qquad\qquad (2.36)$

The elevation of Q above P is found by normal levelling techniques.

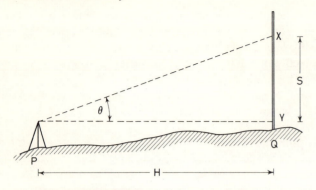

Fig. 2.6. Tangential method—horizontal sights

P. *Instrument station*
Q. *Staff station*
H. *Horizontal distance between survey sta-*
tions P and Q
θ. *Vertical angle subtended at instrument by*
staff intercept
s. *Staff intercept*
X, Y. *Intercepts of central diaphragm line on*
staff obtained from separate pointings of
the telescope

2.3.2   INCLINED SIGHTS

In this case the theodolite is centred over station P, the telescope is
sighted horizontally and the vertical circle reading recorded. Points
Y and X are then sighted on a vertical staff held at station Q, and the
vertical circle readings again observed at both pointings. From these
readings, angles ϕ and θ are known and $(\theta - \phi)$ is the angle which
subtends the staff intercept s, *Fig. 2.7.*
By trigonometry

$$H \tan \theta = XR$$

$$H \tan \phi = YR$$

$\therefore$ $$XR - YR = H(\tan \theta - \tan \phi) = s \qquad (2.37)$$

whence

$$H = \frac{s}{(\tan \theta - \tan \phi)} \qquad (2.38)$$

If the height of the trunnion axis from the ground (*l*) is known, the difference in height (*V*) between the two stations is given by

$$V = l + RQ$$

i.e.

$$V = l + RY - YQ$$

i.e.

$$V = l + H \tan \phi - QY \qquad (2.39)$$

The calculations of $\dfrac{s}{(\tan \theta - \tan \phi)}$ can be somewhat lengthy due

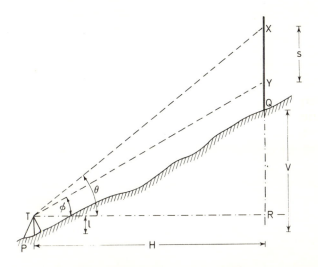

Fig. 2.7. Tangential method—inclined sights
P. Instrument station
Q. Staff station
H. Horizontal distance between stations P and Q
θ. Vertical angle subtended at instrument by line of sight TX with horizontal
φ. Vertical angle subtended at instrument by line of sight TY with horizontal
s. Staff intercept
l. Height of theodolite trunnion axis above station P
T. Trunnion axis of theodolite
V. Vertical distance between stations P and Q
X, Y. Intercepts of central diaphragm line on staff obtained from separate pointings of telescope

to the difference of the two tangents, i.e. $(\tan \theta - \tan \phi)$. Thus the surveyor is faced with two alternatives:

(i) either fix s (say 3 m) and use reciprocal tables, or (ii) fix $(\tan \theta - \tan \phi)$, (say 0·01) and measure s.

The latter method appears to be more popular and quicker from the computational point of view, but it might be less accurate due to interpolation errors in reading the staff.

Various types of instruments have been devised to simplify the field work and the associated office work involved in computations. Among these may be mentioned Eckhold's Omnimeter, the Bell-Elliot tangent-reading tacheometer, the Gradienter, and the Szepessy direct-reading tacheometer. The tangential method, however, is not so commonly used as the ordinary stadia method, and, in general, is not quite so accurate.[4]

2.4　STAVES FOR THEODOLITE TACHEOMETRY

Special tacheometric staves should be used whenever they are available if the best results are to be obtained from theodolite tacheometric methods. If the staff is to be held in the vertical position a levelling bubble should be fitted, to ensure verticality, whereas if it is to be held perpendicularly to the line of sight a special alignment tube should be provided. When ordinary levelling staves are used for tacheometric observations optimum results cannot be expected.

Two parameters which occur repeatedly in theodolite tacheometry are the instrument height (l) and the height of the intercept of the central stadia line above the ground (m), (see *Fig. 2.3*). By making $m = l$, the reduction expressions are considerably simplified. This can be done in two ways, either by sighting an intercept m so that $m = l$, or by using a staff whose index mark can be adjusted to the same height of that of the instrument trunnion axis above the ground. If this index mark is then sighted and brought into coincidence with the central stadia line, only one intercept need be read; this facilitates field work, and considerably simplifies the computation of vertical and horizontal distances.

REFERENCES

1. CLARK. *Plane and Geodetic Surveying for Engineers.* Vol. I. Constable and Co. Ltd. (1923).
2. BENNY. *Mathematics for Students of Engineering and Applied Science.* Oxford University Press (1954).
3. TAYLOR. 'A New Perfectly Anallactic Internally Focussing Telescope.' *Trans. Optical Soc.* **25,** 4 (1923–24).
4. CLENDINNING. *Principles and Use of Surveying Instruments.* Blackie (1966).

CHAPTER 3

Vertical Staff Self-reduction Tacheometers

3.1   INTRODUCTION

Recent developments in tacheometric instruments and techniques have provided means of measuring true horizontal and vertical distances directly, thereby eliminating the necessity of reduction tables and other office computations. Such instruments are now usually referred to as 'self-reduction' tacheometers and sometimes, more briefly, as S.R.T's. Horizontal and vertical angles can also be measured with these instruments to an accuracy of better than 1' minute of arc, and their overall size and weight is similar to that of ordinary theodolites. The use of these instruments is particularly advantageous in detail surveys and can result in considerable economies being effected.

The majority of modern vertical staff tacheometers can be classified as either diagram circle, movable diaphragm or tangential self-reduction types, and typical examples of each will be described in this chapter, together with the special staves which should be used to obtain optimum performance.

3.2   DIAGRAM CIRCLE TACHEOMETERS

In this type of instrument, the stadia interval varies as the telescope is rotated in a vertical plane. This is achieved by marking on glass (diagram) circles, the appropriate reduction curves for distance and elevation or a series of short stadia lines of variable interval. The glass circles are usually vertical and in some instruments the diagram circle and vertical circle are marked on the same glass circle. *Fig. 3.1* shows how the light rays from the staff are deflected through the

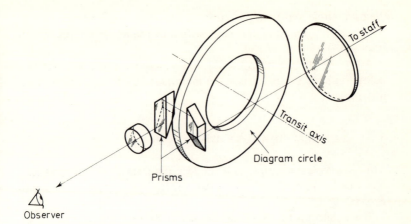

Fig. 3.1. Path of light rays from staff through diagram
circle to observer

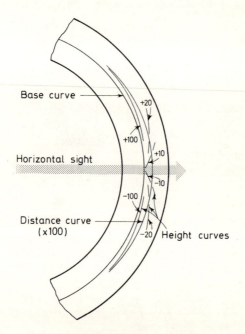

Fig. 3.2. Reduction curves on
the diagram circle of the Zeiss
'Dahlta' tacheometer (Courtesy
of Carl Zeiss (Jena) Co. Ltd.)

diagram circle and then passed into the eyepiece. The observer will then see the image of the staff with the reduction curves superimposed.

A typical instrument is the Zeiss 'Dahlta' 010 which is described below. Table 3.1 summarises the main features of four well-known instruments.

3.2.1   THE ZEISS 'DAHLTA' 010

This is a typical example of a modern diagram circle tacheometer, and the arrangement of the curves on the glass circle is shown in *Fig. 3.2.*

The equations for these curves may be derived as follows:

Let K_H = multiplication constant for horizontal distance.

Let K_v = multiplication constant for vertical distance.

where $K_H = \dfrac{f}{a} = \dfrac{\text{focal length of lens system}}{\text{stadia interval for horizontal distance}}$ $\qquad$ (3.1)

and $\quad K_v = \dfrac{f}{g} = \dfrac{\text{focal length of lens system}}{\text{stadia interval for vertical distance}}$ $\qquad$ (3.2)

Now from equation (2.9) for an anallactic telescope

$$H = K_H \, s \cos^2 \theta \qquad (3.3)$$

∴ from equation (3.1)

$$H = \frac{f}{a} s \cos^2 \theta \qquad (3.4)$$

Now for automatic reduction we have to substitute for s, a staff intercept (s'), which will vary with the angle of elevation (Θ), and which, when multiplied by K_H will give the horizontal distance (H) directly. This can be accomplished by the introduction of a variable stadia interval a' so that from equation (3.4)

$$H = \frac{f}{a'} s' \cos^2 \theta \qquad (3.5)$$

Therefore, in order to obtain the simple relationship

$$H = K_H \, s' \qquad (3.6)$$

Table 3.1 SUMMARY OF TYPICAL DIAGRAM CIRCLE TACHEOMETERS

Instrument	Zeiss 'Dahlta' 010	Fennel F.T.R.A. 'Retac'	Wild RDS	Salmoiraghi 4180
Magnification	25 ×	25 ×	24 ×	28 ×
Objective diameter	40 mm	40 mm	40 mm	40 mm
Shortest focussing distance	3 m	1·7 m	3·3 m	3 m
Graduation of horizontal circle	1', reading by estimation to 6"	1', reading by estimation to 6"	1', reading by estimation to 6"	1', reading by estimation to 6"
Graduation of vertical circle	1', reading by estimation to 12"	1', reading by estimation to 6"	1', reading by estimation to 6"	Tangential scale, estimating to 0·0001
Distance multiplying constant(s) (K_H)	100, 200	100	100	100
Height multiplying constant (K_v)	±10, ±20, ±100	±10, ±20, ±50	±10, ±20, ±50, ±100	H × tangent vertical angle
Accuracy claimed (distance)	±0·10 m in 100 m	±0·10 m in 100 m	±0·10 m in 100 m	±0·10 m in 100 m
Approx. instrument weight	4·9 kg	5·5 kg	6·2 kg	6·7 kg
Remarks	Diagram circle fixed, reduction curves seen only in the face left position of the telescope. Automatic compensating device for correct orientation of diagram circle.	Similar instrument to 'Dahlta'. Unlike the earlier 'Fenta' model. (Whole field of view now available.)	The diagram circle rotates as the telescope is tilted, resulting in gently sloping reduction curves. Fitted with optical plummet.	The diagram circle is marked with straight, variably spaced stadia lines producing constant staff intercepts. Height must be calculated from tangent of vertical angle (read in the eyepiece with staff intercept) and horizontal distance.

the stadia interval must change according to

$$a' = a \cos^2 \theta \qquad (3.7)$$

Substituting for a from equation (3.1)

$$a' = \frac{f}{K_H} \cos^2 \theta \qquad (3.8)$$

From equation (2.10) we have for an anallactic telescope

$$V = K_v s \sin \theta \cos \theta \qquad (3.9)$$

or $\qquad\qquad\qquad V = \tfrac{1}{2} K_v s \sin 2\theta \qquad (3.10)$

Similarly, the variable stadia interval for vertical distance (g') may be deduced as

$$g' = \tfrac{1}{2} \frac{f}{K_v} \sin 2\theta \qquad (3.11)$$

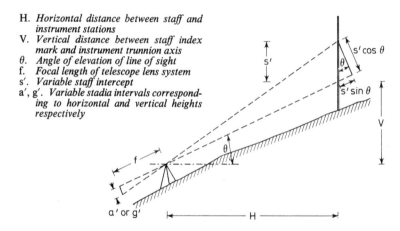

H. *Horizontal distance between staff and instrument stations*
V. *Vertical distance between staff index mark and instrument trunnion axis*
θ. *Angle of elevation of line of sight*
f. *Focal length of telescope lens system*
s'. *Variable staff intercept*
a', g'. *Variable stadia intervals corresponding to horizontal and vertical heights respectively*

Fig. 3.3. *Derivation of the equations for the reduction curves marked on the diagram circle*

Therefore, if the respective stadia interval is varied according to equations (3.8) or (3.11), the horizontal distance and difference in elevation will be obtained directly.

Further simplification can be attained by using a single stadia reading on the staff. This can be accomplished by having an index

mark on the staff. The 'base' curve, etched on the diagram circle, can be brought into coincidence with the index mark, and the horizontal and vertical distances can then be read directly from the intersection, on the staff, of the horizontal and vertical reduction curves of the diagram circle. The staff index mark can usually be adjusted to the same level as the transit axis of the instrument and the difference in ground level between instrument and staff stations obtained directly.

Another point to note is that the staff is not parallel to the plane of the diaphragm plate and this must be taken into account. From *Fig. 3.3* we have

$$\frac{a'}{f} = \frac{s' \cos \theta}{\dfrac{H}{\cos \theta} + s' \sin \theta} \tag{3.12}$$

i.e.

$$a' = \frac{fs' \cos \theta}{\dfrac{H}{\cos \theta} + s' \sin \theta} \tag{3.13}$$

But
$$H = K_H s'$$

∴ substituting in equation (3.13) and rearranging we get

$$a' = \frac{f \cos^2 \theta}{K_H \pm \frac{1}{2} \sin 2\theta} \tag{3.14}$$

The negative sign applies when the staff is below the level of the instrument.

Similarly, from *Fig. 3.3* we have

$$g' = \frac{fs' \cos \theta}{\dfrac{H}{\cos \theta} + s' \sin \theta} \tag{3.15}$$

and
$$V = H \tan \theta$$

∴
$$g' = \frac{fs' \cos \theta}{\dfrac{V}{\sin \theta} + s' \sin \theta} \tag{3.16}$$

but
$$V = K_v s'$$

∴
$$g' = \frac{\frac{1}{2} f \sin 2\theta}{K_v \pm \sin^2 \theta} \tag{3.17}$$

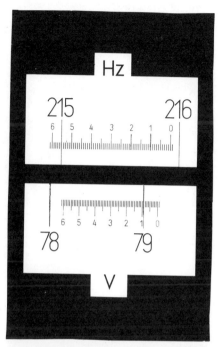

Fig. 3.4 (a).
Horizontal and
vertical circle
readings

Horizontal circle
reading =
215° 55·0′
Vertical circle
reading =
79° 09·0′
(Courtesy of
Carl Zeiss (Jena)
Co. Ltd.)

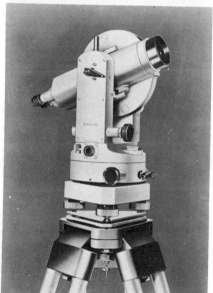

Fig. 3.4 (b). The
Zeiss 'Dahlta' 010
tacheometer
(Courtesy of
Carl Zeiss (Jena)
Co. Ltd.)

Equations (3.14) and (3.17) are the equations of the reduction curves which are etched on the diagram circle.

The Zeiss 'Dahlta' 010 is shown in *Fig. 3.4* together with examples of staff readings and horizontal and vertical circle readings: *Fig. 3.5* shows a diagram of the optical paths inside the instrument.

The multiplication constant for horizontal distance (K_H) is either 100 or 200, and the multiplication constant for vertical distance (K_v) is either ± 10, ± 20, or ± 100, depending upon the vertical angle.

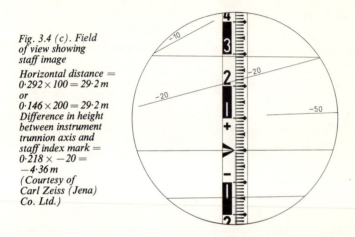

Fig. 3.4 (c). Field of view showing staff image

Horizontal distance =
$0.292 \times 100 = 29.2\,m$
or
$0.146 \times 200 = 29.2\,m$
Difference in height between instrument trunnion axis and staff index mark =
$0.218 \times -20 =$
$-4.36\,m$
(Courtesy of Carl Zeiss (Jena) Co. Ltd.)

The diagram circle is fixed so that distance and height measurements can only be made with the telescope in the face-left position, because only then are the reduction curves visible.

Measurements are carried out in the following way: set up the staff over a ground station and sight the vertical line of the diaphragm on it. With the slow motion screw tilt the telescope until the base curve in the field of view is set upon the index mark of the staff, or if not visible on another decimetre value on the staff, noting the reading in the field book for calculation. The intersection of the distance curve is then read on the staff. (In early diagram circle tacheometers, it was necessary to manually centre the altitude bubble because the reduction curves and the vertical angle divisions were all etched on the same diagram circle, but an automatic compensating device is incorporated in most modern instruments so that fast and error-free distance, height and vertical angle measurements can be made.) The horizontal distance is then calculated from either

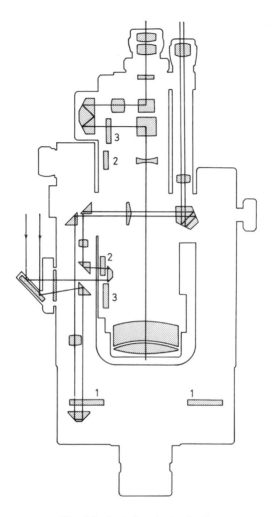

Fig. 3.5. Optical paths in the Zeiss
'Dahlta' 010 tacheometer

1. Horizontal circle
2. Vertical circle
3. Diagram circle
(Courtesy of Carl Zeiss (Jena) Co.
Ltd.)

$$H = 100 \, s_H$$

or
$$H = 200 \, s_H$$

where s_H is the staff intercept between the base curve and the distance curve. In the same way the intersection of the height curve is read on the staff. The difference in height between the telescope trunnion axis and the index mark on the staff is calculated from

$$V = K_V s_V$$

where s_V is the staff intercept between the base curve and the height curve at the vertical diaphragm line. The manufacturers claim an

Fig. 3.6. The Zeiss 'Dahlta' with the Zeiss 'Karti' plotting table attachment (Courtesy of Carl Zeiss (Jena) Co. Ltd.)

accuracy of ± 0.10 m up to a distance of 100 m, and for difference in elevation ± 0.05 m with $K_V = \pm 10$, ± 0.05 m to ± 0.10 m with $K_v = \pm 20$, and ± 0.10 m to ± 0.20 m with $K_v = \pm 100$. The Zeiss 'Dahlta' 010 will measure angles to 1 m and by estimation to 6 s of arc.

A useful attachment to the Zeiss 'Dahlta' 010 is the Zeiss 'Karti'

plotting table shown in *Fig. 3.6*. This enables direct, semi-automatic plotting of traverse surveys and detail filling to be carried out simultaneously, thereby combining the advantages of the plane table technique with self-reduction tacheometry. The plotting area has a diameter of 250 mm and the plotting accuracy for all parts within the mapping area is approximately 0·1 mm. The plotting table rotates with the tacheometer, but the plotting surface remains orientated to the ground by means of a gear system.

3.3 MOVABLE DIAPHRAGM TACHEOMETERS

In the movable diaphragm type of instrument, the stadia interval is altered by a gearing system which operates when the telescope is

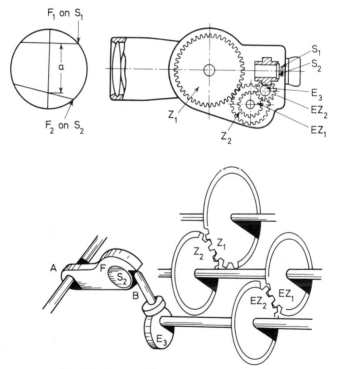

Fig. 3.7. *Gearing and cam mechanism for variation of the stadia interval in the Kern DK–RV vertical staff tacheometer*

tilted. A typical example is the Kern DK–RV self-reduction tacheometer.

3.3.1 THE KERN DK–RV TACHEOMETER

The principle of the Kern DK–RV is as follows. The distance lines
F_1 and F_2 (*Fig. 3.7*) are etched on separate reticules S_1 and S_2 and
the etched surfaces face one another. Reticule S_1 is fixed and S_2 is
movable, being controlled by a mechanism which raises or lowers
the reticule as the inclination of the telescope is varied. The motion
of the reticule is such that the interval 'a' intercepts a length (*s*) on
a staff graduated such that it gives the horizontal distance to the
staff. To avoid estimation of the staff intercept *s* in the vertical

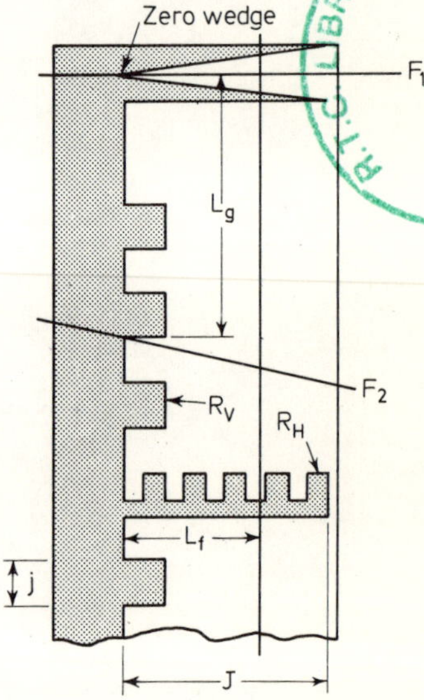

Fig. 3.8. Principle of reading the Kern DK–RV staff

pattern R_v (*Fig. 3.8*) the inclination of the line F_2 is so controlled
by the tilting of the telescope and the gear-cam assembly, that it is
possible to measure *s* in the horizontal pattern R_H. The slope is
determined by the relationship of the intervals *j* and *J*. With line
F_1 set in the zero wedge of the staff and the vertical line moved in
azimuth until line F_2 cuts a metre division of the scale R_v, the staff
intercept is divided into an approximate measure L_g in metres and

a fine measure L_f in decimetres and centimetres. The total horizontal distance is given by the addition of these two intercepts. An example of a staff reading is shown in *Fig. 3.9(a)*.

The variation of the interval 'a' that must be provided by the control mechanism is based upon the following principles. From

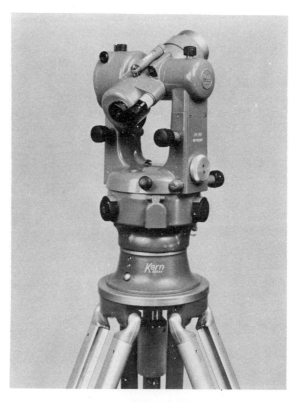

*Fig. 3.9(a). The Kern DK–RV tacheometer
(Courtesy of The Kern Co. Ltd.)*

Fig. 3.10 for a given horizontal distance, *H*, regardless of slope, the instrument must give that distance as the product of a constant, *K*, and the staff intercept s_o or s_θ.

Thus $$H = Ks_o = Ks_\theta \tag{3.18}$$

and so obviously

$$s_o = s_\theta \tag{3.19}$$

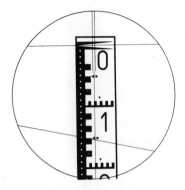

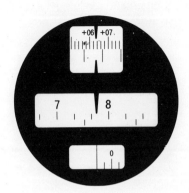

Fig. 3.9(b) (top left). The Kern DK–RV field of view showing staff image

Horizontal distance = 15·32 m

(c) (top right). The Kern DK–RV horizontal and circle readings without micrometer

Horizontal circle reading = 07° 46′
Vertical circle reading (tangent scale) = +0·0634

(d) (left). The Kern DK–RV horizontal and circle readings with micrometer

Horizontal circle reading = 268° 25′ 43″
Vertical circle reading (tangent scale) = +0·0632
(Courtesy of The Kern Co. Ltd.)

Thus the parallactic angle α_θ must be a function of the horizontal parallax α_o and the inclination of the telescope θ. In this way the staff intercept for a given horizontal distance will be constant regardless of the slope.

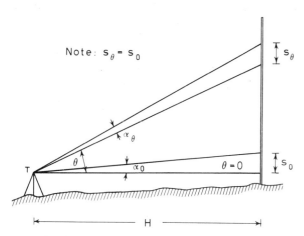

Fig. 3.10. *The principle of variation of the stadia interval in the Kern DK–RV tacheometer*

H. *Horizontal distance between stations*
T. *Trunnion axis of the instrument*
s_θ. *Staff intercept corresponding to parallactic angle α_θ*
s_o. *Staff intercept corresponding to parallactic angle α_o*
θ. *Angle of elevation of telescope*

By inspection in *Fig. 3.10* we obtain

$$s_o = H \tan \alpha_o \qquad (3.20)$$

and from equations (3.18) and (3.20), therefore, we have

$$K = \cot \alpha_o \qquad (3.21)$$

also from *Fig. 3.10* we have

$$s_\theta = H(\tan (\theta + \alpha_\theta) - \tan \theta) \qquad (3.22)$$

Now:

$$\tan(\theta + \alpha_\theta) - \tan \theta = \frac{\tan \theta + \tan \alpha_\theta}{1 - \tan \theta \tan \alpha_\theta} - \tan \theta_\theta \qquad (3.23)$$

$$= \frac{\dfrac{\sin\theta}{\cos\theta} + \tan\alpha_\theta}{1 - \dfrac{\sin\theta}{\cos\theta}\tan\alpha_\theta} - \frac{\sin\theta}{\cos\theta}$$

$$= \frac{\cos\theta\left\{\dfrac{\sin\theta}{\cos\theta} + \tan\alpha_\theta\right\} - \sin\theta\left\{1 - \dfrac{\sin\theta}{\cos\theta}\tan\alpha_\theta\right\}}{(1 - \dfrac{\sin\theta}{\cos\theta}\tan\alpha_\theta)\cos\theta} \tag{3.24}$$

$$= \frac{\sin\theta + \tan\alpha_\theta\cos\theta - \sin\theta + \dfrac{\sin^2\theta}{\cos\theta}\tan\alpha_\theta}{\cos\theta - \sin\theta\tan\alpha_\theta} \tag{3.25}$$

$$= \frac{\tan\alpha_\theta(\cos^2\theta + \sin^2\theta)}{\cos^2\theta - \tan\alpha_\theta\sin\theta\cos\theta} \tag{3.26}$$

Thus from equations (3.22) and (3.26)

$$s_\theta = \frac{H\tan\alpha_\theta}{\cos^2\theta - \tan\alpha_\theta\sin\theta\cos\theta} \tag{3.27}$$

Substituting equations (3.20) and (3.27) in equation (3.18) we have

$$KH\tan\alpha_o = \frac{KH\tan\alpha_\theta}{\cos^2\theta - \tan\alpha_\theta\sin\theta\cos\theta} \tag{3.28}$$

$\therefore \qquad \tan\alpha_o\cos^2\theta - \tan\alpha_o\tan\alpha_\theta\sin\theta\cos\theta = \tan\alpha_\theta \tag{3.29}$

$\therefore$ dividing through by $\tan\alpha_o$

$$\frac{\tan\alpha_\theta}{\tan\alpha_o} = \cos^2\theta - \tan\alpha_\theta\sin\theta\cos\theta \tag{3.30}$$

$\therefore \qquad \cos^2\theta = \dfrac{\tan\alpha_\theta}{\tan\alpha_o} + \tan\alpha_\theta\sin\theta\cos\theta \tag{3.31}$

$\therefore \qquad \cos^2\theta = \tan\alpha_\theta\,(\cot\alpha_o + \sin\theta\cos\theta) \tag{3.32}$

$$\tan\alpha_\theta = \frac{\cos^2\theta}{\cot\alpha_o + \sin\theta\cos\theta} \tag{3.33}$$

But from equation (3.21) $K = \cot\alpha_o$

$\therefore \qquad\qquad \tan\alpha_\theta = \dfrac{\cos^2\theta}{K + \frac{1}{2}\sin 2\theta} \tag{3.34}$

Equation (3.34) is the function which the gearing and cam controlling the moving reticule must follow. A further point to note is that the projection of interval j of the vertical scale R_v shortens as the inclination of the telescope (θ) increases, whereas the interval J of the horizontal scale R_H is independent of θ. Therefore the obliquity of the sloping reticule line must vary with θ.

Fig. 3.11 demonstrates the method by which this is done. When

then
$$\theta = O \tan \beta_o = \frac{j}{J} \cos \alpha_o \simeq \frac{j}{J} \ldots \ldots \tag{3.35}$$

and when
$$\theta \neq O \tan \beta_\theta = \tan \beta_o \cos (\theta + \alpha_\theta) \tag{3.36}$$

The controlling device of the stadia interval 'a' is activated by gear wheels Z_1 and Z_2, the offset pair EZ_1 and EZ_2 and cam E_3

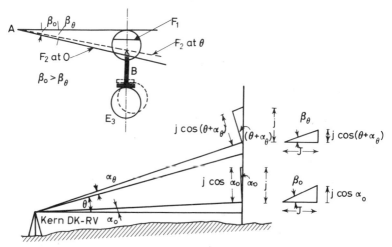

Fig. 3.11. The variation of the obliquity of the sloping
stadia line in the Kern DK–RV tacheometer
The size of j has been exaggerated

(*Fig. 3.7*). The cam not only imparts translation to the reticule S_2, but also rotation since the reticule pivots about point A. Thus the cam also varies the slope β of line F_2.

The Kern DK–RV is shown in *Fig. 3.9* together with examples of staff and horizontal and vertical circle readings. The manufacturers claim an accuracy of ± 0.03 to ± 0.05 m per 100 m for this instru-

ment. The staff is so constructed that the index mark can be set to the same level as the trunnion axis of the instrument. The vertical circle carries a tangential graduation and the difference in elevation is obtained by multiplying the reduced horizontal distance by the tangent of the vertical angle. This instrument also measures horizontal angles direct to 10 sec and by estimation to 1 sec of arc.

3.3.2 THE KERN CENTERING TRIPOD

A successful attempt to reduce the time required for centering and levelling instruments over a ground station is represented by the Kern tripod system which is now standard for all of their instruments

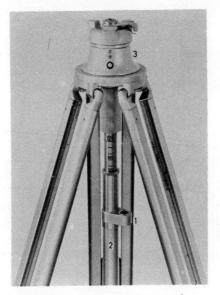

1. *Circular levelling bubble.*
2. *Centering rod.*
3. *Movable centering and levelling base.*

Fig. 3.12. The Kern centering tripod
(Courtesy of The Kern Co. Ltd.)

(*Fig. 3.12*). This system automatically establishes coarse levelling of the tripod top when centering over the station with the centering rod. Precise levelling of the instrument can then be obtained by the use of special levelling cams, which replace the conventional levelling screws. The cams can only be rotated about 180°, this being sufficient to level the instrument under all circumstances.

In addition to speeding up the whole operation of centering and levelling, the system has another advantage which is particularly useful in tacheometric and trigonometrical levelling, in that the height of the trunnion axis of the telescope above the base of the instrument remains constant and therefore can be read directly from the centering rod. This is not possible with conventional levelling screws as the trunnion axis height above the base plate will vary and therefore cannot be read directly from the centering rod.

Richardus[1] estimates the accuracy of centering rods to be approximately 3 mm; Mathias[2] claims plumbing over a station is accurate within approximately 1 mm for the Kern system which also levels the tripod top to about 1 or 2 min of arc. In this way the surveyor can set up quickly and easily on all kinds of terrain, and a substantial increase in the number of points per hour surveyed can be obtained. The manufacturers claim that compared with the conventional tripod, set-up time is cut by at least 50% and where the terrain is difficult or the observers inexperienced, a much greater time saving can be realised. Similar advantages are also obtained with the Zeiss (Oberkochen) ball base and RTa4 self-reduction tacheometer.

3.4   TANGENTIAL SELF-REDUCTION TACHEOMETERS

There are not many instruments manufactured today that employ a vertical staff and the principles of tangential reduction, probably because the diagram circle and movable diaphragm types of tacheometers have been found easier and more convenient to use.

The principle involves two staff sightings which are normally chosen to coincide with specific vertical angle tangent readings. With the knowledge of the staff intersections (X and Y) and the tangent readings (p_1 and p_2) the horizontal distance can be computed (*Fig. 2.7*) and from this, the vertical distance also can be calculated.

3.4.1   THE SZEPESSY TACHEOMETER THEODOLITE

The Szepessy Tacheometer serves to illustrate the principles of tangential reduction. The instrument will measure both horizontal and vertical distances within the angular elevation and depression range from $+40°$ to $-40°$ and the tangent division scale, seen in the

telescopic field of view, is numbered both positively and negatively from 0 to 80.

Both horizontal and vertical distances are obtained by sighting twice at the staff so that the image of each tangent division line should fall between the double diaphragm line of the vertical circle reading microscope. If the staff readings are X and Y and the respective tangent readings p_1 and p_2 we have

$$s = X - Y$$

The stadia ratio $\qquad K = \dfrac{100}{p_1 - p_2}$, and

Horizontal distance $\qquad H = Ks$

The difference in elevation between the staff station and the trunnion axis of the instrument is

$$V = \frac{H}{100}p_2 - Y \text{ or } V = \frac{H}{100}p_1 - X$$

The instrument is suitable for minor triangulation and topo-graphical surveys, contouring, levelling, and traverse surveys in urban areas. It may be argued that the Szepessy instrument is not truly self-reducing because of the necessity of office computations.

3.5 ERRORS IN VERTICAL STAFF TACHEOMETRY

The most important sources of error in vertical staff tacheometry may be divided into three classes:

1. Errors in reading the staff intercept.
2. Instrumental errors.
3. Errors due to natural causes.

The accuracy of reading the staff intercept depends upon many factors including the following: length of sight, definition and magnifying power of the telescope, fineness of the diaphragm lines, elimination of parallax, clearness of the atmosphere, focusing, graduation and positioning of the staff, and the stability of the instrument and staff. In careful tacheometric work the staff must be straight and be held in a truly vertical position. The use of special staves, fitted with spot-levels to ensure verticality, is recommended. Any

possible bending of the staff must be eliminated as this would result in an increase in the staff intercept and consequently an increase in the reduced horizontal and vertical distances. For example, a staff bent so the height of the camber is 20 mm would result in an error of 21 mm in a horizontal sight of 400 m.

Measured results are greatly falsified by any deviation of the staff from the vertical position. As the inclination of the line of sight increases, the errors also increase, and, therefore, the staff should always be carefully erected on steep ground. The error can be positive or negative depending on the vertical angle and the direction of the inclination of the staff. Table 3.2 lists the results evaluated by

Table 3.2   FRACTIONAL ERRORS DUE TO NON-VERTICALITY OF STAFF

Vertical angle	Positive error			Negative error		
	Tilt of staff			Tilt of staff		
Θ	10'	1°	2°	10'	1°	2°
3°	$\frac{1}{6240}$	$\frac{1}{930}$	$\frac{1}{410}$	$\frac{1}{6630}$	$\frac{1}{1320}$	$\frac{1}{820}$
5°	$\frac{1}{3480}$	$\frac{1}{595}$	$\frac{1}{270}$	$\frac{1}{4000}$	$\frac{1}{730}$	$\frac{1}{410}$
10°	$\frac{1}{1920}$	$\frac{1}{310}$	$\frac{1}{150}$	$\frac{1}{1960}$	$\frac{1}{340}$	$\frac{1}{180}$
15°	$\frac{1}{1280}$	$\frac{1}{205}$	$\frac{1}{100}$	$\frac{1}{1300}$	$\frac{1}{220}$	$\frac{1}{115}$
20°	$\frac{1}{940}$	$\frac{1}{155}$	$\frac{1}{75}$	$\frac{1}{950}$	$\frac{1}{160}$	$\frac{1}{85}$
25°	$\frac{1}{735}$	$\frac{1}{120}$	$\frac{1}{60}$	$\frac{1}{740}$	$\frac{1}{125}$	$\frac{1}{65}$
30°	$\frac{1}{590}$	$\frac{1}{98}$	$\frac{1}{47}$	$\frac{1}{600}$	$\frac{1}{100}$	$\frac{1}{52}$

Redmond[3] for the fractional error due to non-verticality of a 12-ft staff. It is also important that the plane of the staff graduations is normal to the line of sight.

Many of the factors listed as affecting the accuracy of reading the staff intercept will depend upon the optical quality of the telescope. The aperture of the object glass and the degree of magnification will control the ability of the telescope to reveal detail to the eye. An objective of 40 mm can clearly resolve two points separated by 3 sec of arc, whereas the unaided eye can resolve an angle of about 1 min and, therefore, a magnification of at least 20 × is required before the eye can see all that the objective glass has to disclose in the image. Most tacheometers have objective glass diameters of 40 mm and magnifications ranging from 24 × to 28 ×. The general development and improvement in the design and quality of the modern tacheometric telescope has now reached a very